EXERCISE ONE

1. Name the tissues whose cells are thickened with:
    a) Cellulose and pectin.

    b) Lignin.

2. The diagram below represents a fern.

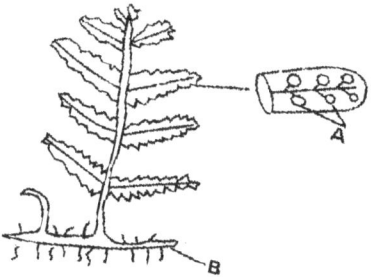

(a) Name Parts labeled A and B.

(b) To which division does the plant belong?

3. State three measures that can be taken to control infection of man by protozoan parasites

4. Explain how the following factors hinder self pollination in plants:
    (i) Protogyny

    (ii) Dioecism

5. Explain the likely effect on humans and other organisms of untreated sewage discharged into water body that supplies water for domestic use.

6. Name two structures in herbaceous stems that enhance their support.

7. a) Define the term immunity.

    b) Distinguish between natural immunity and acquired immunity.

    c) Identify one immunizable disease in a country.

1

8. State three differences between osmosis and active transport.

9. The diagram below illustrates part of a nephron from a mammalian kidney.

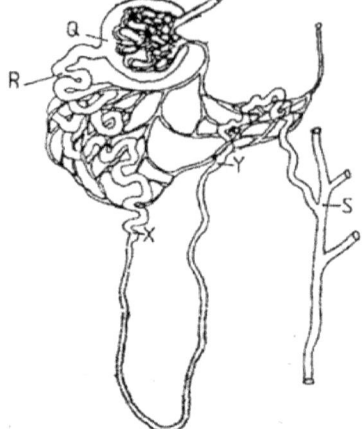

a) Name the fluid found in the part labeled Q.

b) Identify the process responsible for the formation of the fluid named in (a) above.

c) Which two hormones exert their effect in the nephron?

10. State three characteristics of members of kingdom Monera that are not found in other kingdoms.

11. What is meant by the following biological terms?

     i)   Crenation

     ii)   Haemolysis

     iii)   Plasmolysis

12. The diagram below shows a stage during fertilization in flowering plant.

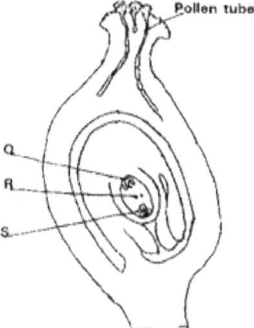

a)   Name the parts labeled Q, R, and S.

b)   State the function of the pollen tube.

13. a) State the major factor in the 'Global warming' experienced in the world today.

b) Suggest two ways of reducing the Global warming.

14. An experiment was set to investigate a certain aspect of response. A seedling was put on a horizontal position as shown in figure M below. After 24 hours, the set up was as shown in figure N.

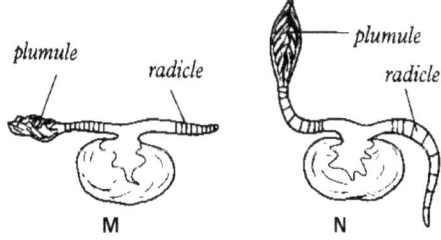

a) Name the response exhibited.
b) Explain the curvature of the shoot upwards.

15.   The paddles of whales and the fins of fish adapt these organisms to aquatic habitats.
       a) Name the evolutionary process that may have given rise to these structures.

       b) What is the name given to such structures?

c) Give two examples of vestigial organs in man.

16. a) Name a protein and vitamin involved in blood clotting.
     i) Protein.
     ii) Vitamin

  b) Explain why blood is not normally used for transfusion after one month.

17. A group of Form four students set up an experiment to investigate a biological process using termites. They used a small box in which a portion was covered with black paper and had moist soil. The open part had dry soil. Termites were placed inside in open area of the box.

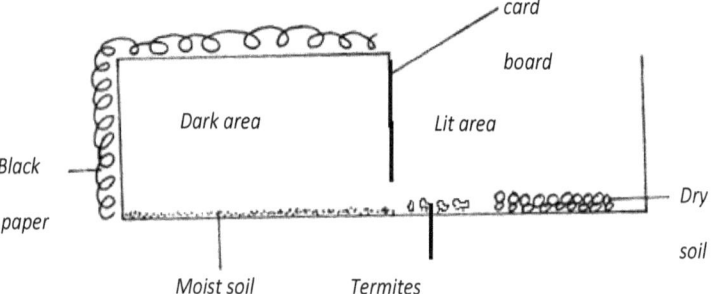

a) Predict what happened to the termites after 30 minutes.

b) What form of response is exhibited by termites?

c) State one biological significance of the above response to termites.

18. a) Name two fins in a bony fish which perform the following functions:-
    i)   Changing direction.

    ii)   Control pitching.

  (b) State the role of myotomes in fish.

19. The diagram below represents an experimental set up to investigate a certain scientific concept. The potted plant was first destarched by keeping it in dark for four days.

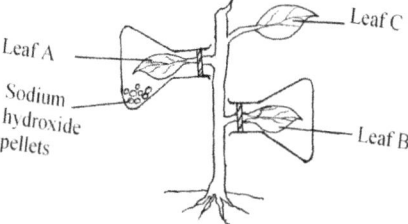

The set up was then placed in sunlight for five hours and leaves were tested for starch.

a)   What scientific concept was being investigated?

b)   i)   Give the results likely to be obtained after starch test for A and B.

        ii)   Account for the results in leaf A in b (i) above.

        c) Why was leaf C included in the set-up?

20. a) Explain the importance of transport in plants.

    b) What is the role of root hairs in plants?

21. a) Identify the source of urea that is removed via the kidneys in a healthy human being.

    b) Explain why a pregnant woman excretes less urea compared to a woman who is non-pregnant.

22. Study the reaction below and answer the questions that follow.

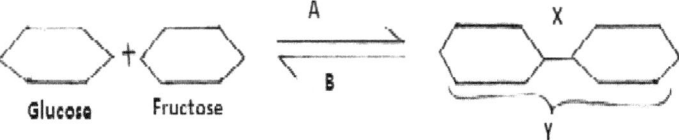

    a) What biological processes are represented by A and B?

b) Identify the product Y.

c) State the bond represented by X.

23. .Explain the events of the light stage of photosynthesis.

24. Explain what happens in humans when the concentration of glucose in the blood rises above the normal level.

25. a) Outline the main features of Lamarckian theory of evolution.

b) In view of modern genetics, explain why Lamarck's theory is unacceptable.

c) Name one factor in nature that increases the process of evolution.

ANSWERS TO EXERCISE ONE

1. Name the tissues whose cells are thickened with:
   d) Cellulose and pectin.
      Collenchyma;
   e) Lignin.
      Sclerenchyma;

2. The diagram below represents a fern.

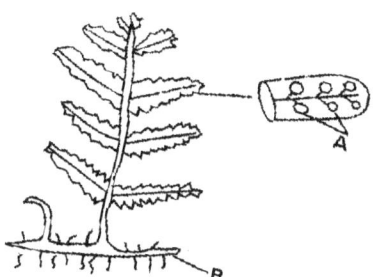

(a) Name Parts labeled A and B.
   A – *Sori;*   rej. *sorus*

   B – *Rhizome;*

(b) To which division does the plant belong?
   Pteridophyta;

3. State three measures that can be taken to control infection of man by protozoan parasites

   Improving sanitation/hygiene; using insecticides to kill vectors; avoiding

   indiscriminate sexual intercourse; clearing bushes/tall grass around the house;

   draining stagnant water; proper disposal of household refuse;

4. Explain how the following factors hinder self pollination in plants:
   (i) Protogyny
   Stigma matures earlier and is ready to receive pollen grains before the anthers are

   ready;

   (ii) Dioecism
      Male and female gametes occur in separate plants;

5. Explain the likely effect on humans and other organisms of untreated sewage discharged into water body that supplies water for domestic use.

7

Contains disease – causing micro-organisms which may cause outbreak of water borne diseases; faecal material is broken down by saprophytes leading to depletion of dissolved oxygen thus suffocation of aquatic organisms; breakdown of matter releases nutrients which enrich the water resulting in eutrophication;

6. Name two structures in herbaceous stems that enhance their support.

   Possession of hooked spines; tendrils; twining stems; adventitious roots;

7. a) Define the term immunity.

   Ability of the body to identify/ recognize foreign antigens and develop mechanisms of destroying them / ability to resist infection;

   b) Distinguish between natural immunity and acquired immunity.

   Natural immunity is inborn /inherited /passed from parents to offspring while acquired immunity is obtained in life;

   c) Identify one immunizable disease a country.

   Tuberculosis; poliomyelitis; diphtheria; whooping cough; measles;

8. State three differences between osmosis and active transport.

   Osmosis involves movement of water /solvent molecules, active transport involves movement of solute molecules; osmosis does not require energy, active transport requires energy; in osmosis molecules move along a concentration gradient, in active transport molecules move against a concentration gradient;

9. The diagram below illustrates part of a nephron from a mammalian kidney.

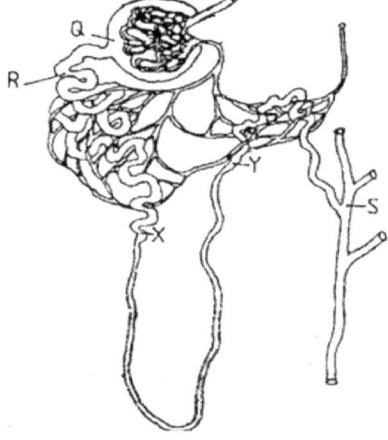

a) Name the fluid found in the part labeled Q.

   Glomerular filtrate;

b) Identify the process responsible for the formation of the fluid named in (a) above.

   Ultra-filtration / pressure filtration;

c) Which two hormones exert their effect in the nephron?
   Antidiuretic hormone / vasopressin; Aldosterone;

10. State three characteristics of members of kingdom Monera that are not found in other kingdoms.

Nucleus lack nuclear membrane / organelles not membrane bound; nucleus not organized; mitochondria absent / most organelles absent; cell wall made of mucoprotein;

11. What is meant by the following biological terms?

   ii)   Crenation

         Shrinking of red blood cells/ animal cells as a result of water loss by osmosis (when placed in hypertonic solution);

   ii)   Haemolysis

         Bursting of red blood cells as a result of uptake of water by osmosis (when placed in hypotonic solution);

   iii)   Plasmolysis

         Shrinking and pulling away of the cell membrane from the cell wall of plant as a result of water loss by osmosis;

12. The diagram below shows a stage during fertilization in flowering plant.

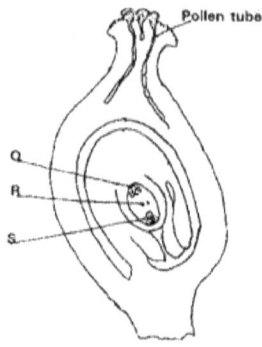

Pollen tube

Q

R

S

a) Name the parts labeled Q, R, and S.

    Q – Antipodal cell(s);

    R – Polar nucleus / body;

    S – Functional egg cell;

b) State the function of the pollen tube.
pathway through which male nuclei reach the embryo sac / improves efficiency of fertilization; its tip produce lytic enzyme which dissolves the embyo sac wall to allow entry of male nuclei;

13. a) State the major factor in the 'Global warming' experienced in the world today.

    Carbon (IV) Oxide;   *rej.Carbon (iv) Oxide*

b) Suggest two ways of reducing the Global warming.

    Reducing use wood / fossil fuels; planting more trees / afforestation or re-afforestation;

14. An experiment was set to investigate a certain aspect of response. A seedling was put on a horizontal position as shown in figure M below. After 24 hours, the set up was as shown in figure N.

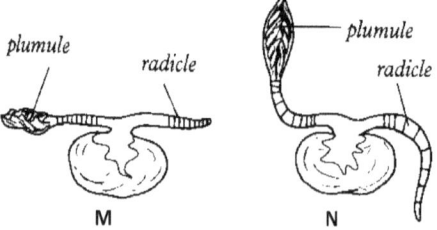

plumule   radicle   plumule   radicle

M   N

a) Name the response exhibited.
    Geotropism;

b) Explain the curvature of the shoot upwards.

Gravity causes high concentration of auxins on the lower part of the shoot; this causes faster elongation of cells on the lower part compared to the upper part; making the shoot to curve upwards;

15. The paddles of whales and the fins of fish adapt these organisms to aquatic habitats.
    a) Name the evolutionary process that may have given rise to these structures.

    Convergent evolution;

    b) What is the name given to such structures?

    Analogous structures;
    c) Give two examples of vestigial organs in man.

    Coccyx; appendix;

16. a) Name a protein and vitamin involved in blood clotting.
    i) Protein.
    Fibrinogen;

    ii) Vitamin
    (Vitamin) K;

    b) Explain why blood is not normally used for transfusion after one month.

    Most of the red blood cells will have died;

17. A group of Form four students set up an experiment to investigate a biological process using termites. They used a small box in which a portion was covered with black paper and had moist soil. The open part had dry soil. Termites were placed inside in open area of the box.

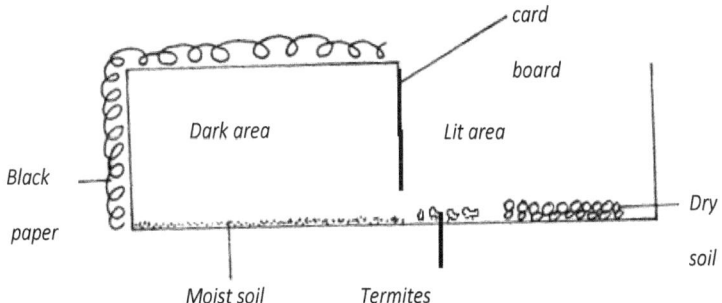

a) Predict what happened to the termites after 30 minutes.

Moved to the dark area;

b) What form of response is exhibited by termites?

Negative phototaxis / positive hydrotaxis

      a) State one biological significance of the above response to termites.

      To escape predation; to reduce dessication;

18. a) Name two fins in a bony fish which perform the following functions:-

    iii)   Changing direction.

       Pectoral fins;

    iv)   Control pitching.

       Pectoral and pelvic fins;

  (b) State the role of myotomes in fish.

  Contract and relax alternately to bring about undulating movement;

19. The diagram below represents an experimental set up to investigate a certain scientific

concept. The potted plant was first destarched by keeping it in dark for four days.

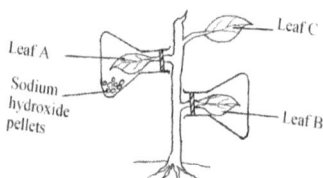

The set up was then placed in sunlight for five hours and leaves were tested for starch.

a)   What scientific concept was being investigated?

     Photosynthesis;

b)   i)   Give the results likely to be obtained after starch test for A and B.

       A and B.

    A – Negative test / starch absent;

    B – Positive test / starch present;

    ii)   Account for the results in leaf A in b (i) above.

    Sodium hydroxide absorbed all the Carbon (IV) Oxide hence no photosynthesis;

  c) Why was leaf C included in the set-up?

    Control experiment;

20. a) Explain the importance of transport in plants.

  Supplies water and mineral ions to the (photosynthetic) cells; conduct products of

  photosynthesis / nutrients to all parts of the plant / translocation;

b) What is the role of root hairs in plants?

Absorption of water and mineral ions from the soil;

21. a) Identify the source of urea that is removed via the kidneys in a healthy human being.

Deamination of excess proteins / amino acids in the liver;

b) Explain why a pregnant woman excretes less urea compared to a woman who is non-pregnant.

Amino acids are used in the formation of foetal tissues; thus has less excess to be eliminated;

22. Study the reaction below and answer the questions that follow.

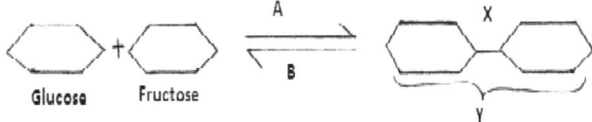

a) What biological processes are represented by A and B?

A – *Condensation;* B – *Hydrolysis;*

b) Identify the product Y.

Sucrose;

c) State the bond represented by X. *Glycosidic;*

23. .Explain the events of the light stage of photosynthesis.

Light energy is s absorbed by chlorophyll molecules; used to split water molecule into oxygen and hydrogen atoms/ ions; light energy is converted into chemical energy (ATP) and stored;

24. Explain what happens in humans when the concentration of glucose in the blood rises above the normal level.

Insulin is produced which increases oxidation of glucose; facilitate conversion of glucose into glycogen / fats for storage; inhibits conversion of glycogen into glucose;

25. a) Outline the main features of Lamarckian theory of evolution.

Use and disuse of structures / when structures are not used for a long time they shrink and when used they develop properly; transmission of physically acquired characteristics / physically acquired characteristics are passed on to the offspring;

b) In view of modern genetics, explain why Lamarck's theory is unacceptable.

phynotypically / physically acquired characteristics which do not affect the genes cannot be inherited;

c) Name one factor in nature that increases the process of evolution.

Natural selection; cross- breeding; mutation;

EXERCISE TWO

1. The diagram below shows a life cycle of a cockroach

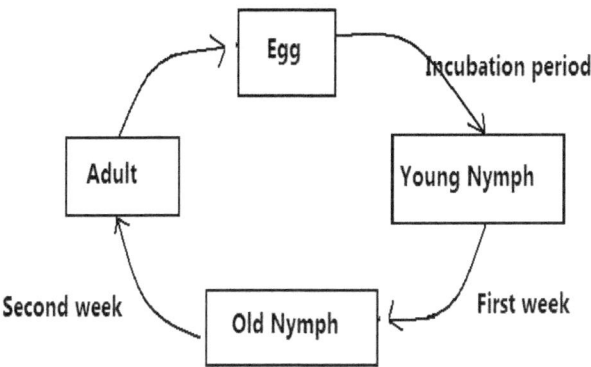

    (a) **Name** the hormone that would be at high concentration during the first and second week and their functions.

(i) First week

(ii) Second week

    (b) **Name** the structure that produces hormone named in a (ii) above

c) Name the process represented by the life cycle above

d) State two importance for the process named in (c) above

2. The diagram below represents part of a geasous system in a grasshopper.

      P   Q

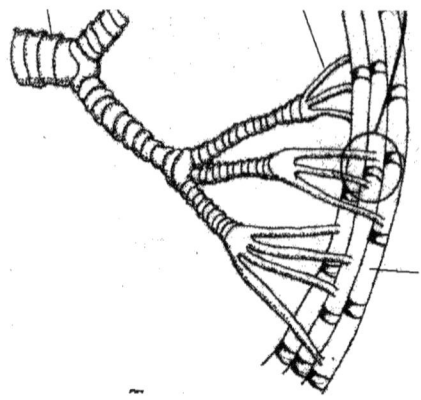

a) Name the structures labeled P and Q

b) State the function of the structure labeled P

c) Describe the path taken by carbon (IV) oxide from the tissues of the insect the atmosphere

d) How is the structure labeled Q adapted to its functions

3. The set up below illustrates an experiment to demonstrate a certain biological process, before the addition of the yeast suspension the glucose solution was first boiled and then cooled at 40°C.

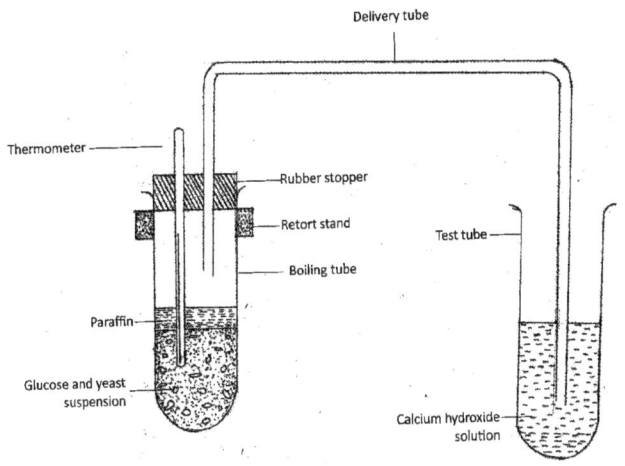

a) What was the aim of the experiment?

b) What observations would you make in the tubes a few minutes after the experiment begun

c) Explain the observations made in (b) above

d) Why was glucose solution boiled before cooling at 40°C

e) How can you set up a control experiment for the above

4. The following are short messages (sms) on cell phone communication between Mrs. Mkenzie and her husband. They can be used as analogies of gene mutation

	Intentended message	Actual message
1.	I want a drive	I want a driver
2.	Yesterday was my shopping day	Yesterday was my hopping day

3	My skirt was stolen	My shirt was stolen
4	Tommorrow I will be visiting my team	Tommorow I will be visitng my mate

a) For each of these messages identify the type of gene mutation illustrated

b) State one example of chromosomal mutation that lead to

i) Change in chromosome structure

ii) Change in chromosomal number

c) Explain why genetic counseling is termed as one practical application of genetics

5. The following is a photograph of s dissected mammal. Study the photograph and answer the questions that follows

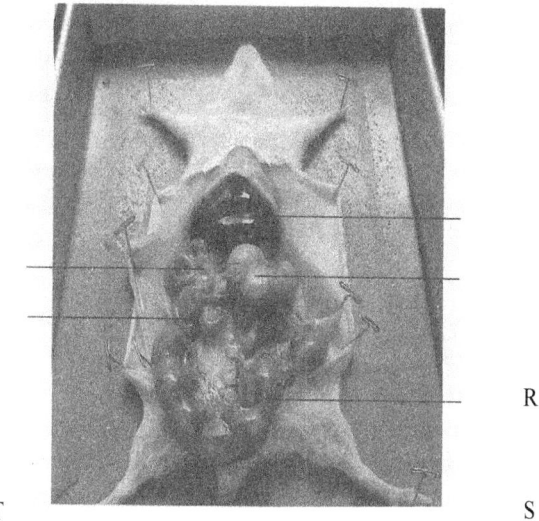

R

S

T

a) Name the structures labeled R,S and T

b) On the photograph, label and name the site of production of vitamin K

c) State one function of the following parts :- S and T.

d) i) State the sex of the dissected mammal

ii) Give a reason for your answer in d (i) above

6. The figure below shows the changes in the concentration of various substances in a river following the discharge of untreated sewage into it. Study it and answer the questions that follow

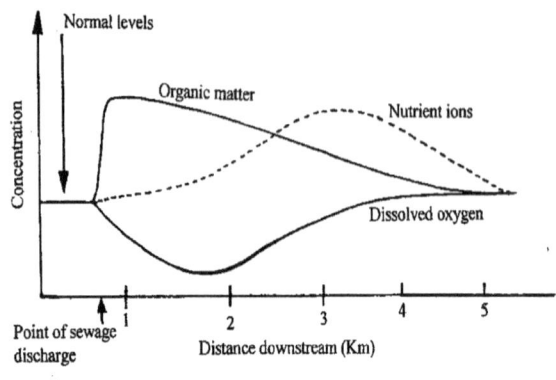

a) Account for the changes in the concentration of:

i) Organic matter

ii) Nutrient ions

iii) Dissolved oxygen

b) Describe the changes you would expect to observe with respect to:

i) Fish population

ii) Water plants and photosynthetic algae

c) State four ways of controlling the type of pollution illustrated above

7. Describe how the following types of plants are adapted to their habitats:

a) Mesophytes

b) Halophytes

c) Hydrophytes

8. Discuss the adaptations of the human eye to its functions

ANSWERS TO EXERCISE TWO

1. (a)　(i) Juvenile hormone

Function – forms larval cuticle/ inhibits moulting metamorphic effects of hormone in the larval stage

(ii) Ecdysone / moulting hormone

Function – moulting to allow growth and metamorphosis

(b)　Prothoracic gland

(c)　Metamorphosis/ incomplete metamorphosis

(d)　i) It reduces competition for food since they feed on different food substances

　　ii) Adapts the organism to escape adverse environmental conditions

2. a) P – rings of chitin/ spiral bands

　Q – tracheoles

b) Circular rings keep the trachea open when pressure is exerted inside the lumen is low

c) Carbon (IV) oxide is of high concentration in the muscle tissue therefore it diffuses from the tissues into the tracheoles and moves to the trachea and out through the spiracles into the atmosphere

d) – the walls are thin and lack chitin for rapid diffusion of gases

　- Walls are moist to dissolve gases

　- They are highly branched to increase the surface area for gaseous exchange

3. a) –To show that carbon (IV) oxide is produced during anaerobic respiration.

b) –There is production of bubbles in the boiling tube of carbon (IV) oxide leading to the formation of a white precipitate in the test tube containing calcium hydroxide solution.

c) – Yeast cells undergo anaerobic respiration prodicing carbon (IV0 oxide gas which dissolves in calcium hydroxide solution forming a white precipitate; production of the gas results in bubbles

d) – To kill other micro-organisms present in glucose solution thus eliminating microbial respiration

e) – Through the use of glucose solution without yeast cells

4.   a) 1- Insertion

      2 – Deletion

      3 – Substitution

      4 – Invertion

b) i) Deletion, Duplication

ii) Non-disjuction

c) It involves provision of information and advice on genetically inherited disorders, their risks and outcomes

5.   a) R- Liver

 S – Stomach

 T – Ileum

b) To label colon on the photograph

c) S – Site of initial protein digestion/ site for temporary storage of food

 T – Site of completion of digestion and absorption

d) i) – Female

 ii) – Presence of uterus

-    Presence of foetus in uterus

6. a) i) Organic matter stimulates proliferation of saprophytic bacteria; that breakdown organic matter into soluble materials; thus a steady decrease in the amount of organic matter in the water downstream

ii) The breakdown of organic matter by saprophytic bacteria releases nutrient ions e.g. sulphates and phosphates; increasing their concentration in the water

iii) Saprophytic bacteria use up dissolved oxygen as they respire; decreasing the concentration of oxygen; as the level of organic matter declines downstream so does the activity of saprophytic bacteria; oxygen from air dissolves in the water returning its concentration to normal

b) i) Between the point of sewage discharge and the point where organic matter returns to normal levels; fish population decreases drastically; this is due to lack of sufficient oxygen which the fish needs for respiration

ii) About one and half km downstream from the point of sewage discharge; water plants and algea proliferation; this occurs due to the large number of nutrient ions released from the breakdown of organic matter by saprophytic bacteria; their numbers get back to normal downstream as the nutrients get exhausted

c) – treatment of domestic waste before discharging into water bodies

- using biotechnology to treat waste

- banning use of phosphate based detergents

- replacing lead pipes with plactic pipes

7. a) Mesophytes

- Trees may grow very tall in forests ecosystem due to competition for light as vegetation is very dense

- Some plants are climbers which support themselves on large tree in an attempt to reach light

- Some plants are epiphytes growing on tree branches to reach light

- Some undergrowth plants have numerous chloroplasts which are sensitive to low light intensity to enable them carry out photosynthesis in low light intensity

- Many plants have leaf mosaic that minimise overlapping and overshadowing and increase exposure of leaves to light

- those in areas with adequate water supply posses broad leaves with thin cuticle and many stomata on both sides of the leaf to increase transpiration
- Those in dryer areas have fewer stomata w3hich are mainly located on the lower surface to reduce transpiration.

- Some which leave in wet areas have shallow roots to absorb less water

- Large tall trees have developed butress roots or prop roots for extra support.

- Those in dryer areas have deep roots to absorb water from water table

- Some have waxy and glossy surface to reflect light to reduce absorption of light hence reduce transpiration also to drip off rain water.

b) Halophytes

- They have roots that concentrate a lot of salts in their cells by active transport; to enable them off set osmotic imbalance and take in water by osmosis

- Some have salt glands that secretes excess salts

- Some have water storage tissues to store water that has been taken in.

- Some like mangrooves have pneumatophores which have lenticels for gaseous exchange

- Some mangrooves have stilt roots for extra anchorage in mudflats.

- Most halophytes are found growing close to the water surface to enable them get sufficient light for photosynthesis

- Those in deeper water have highly sensitive chloroplasts to photosynthesise under low light intensity

- Some e.g. coconut have fruits with large aerenchyma tissue to enable them float.

c) Hydrophytes

- Most emertgent and floating types have broad leaves with many stomata on upper surface to provide a large surface area for gaseous exchange

- Some submerged hydrophytes have leaves which are deeply dissected into thread- like straws to provide a large surface for absorption of maximum light for photosynthesis

- Some submerged hydrophytes have leaves with numerous and sensitive chloroplasts that synthesise under low light intensity

- Many hydrophytes have aerenchyma tissues filled with air to enable them float and store gases for gaseous exchange

- They have poorly developed roots that lack root hairs to reduce absorption of water

- Their flowers are raised above the water level to allow for pollination for submerged and emergent species

8. -The **sclerotic** layer which contains tough connective tissue fibres which helps it to support and protect the other parts of the eye ball.

-The **choroids** which contain many blood capillaries which supply oxygen and nutrients of the retina and removes metabolic wastes from eye.

-Its highly pigmented, to prevent reflection of light within the posterior chamber of the eye ball.

-The **retina** which contains photoreceptor cells called cones and rods. It is said to be the light sensitive part of the eye. Cones are adapted for light and colour vision while rods are adapted for dim light vision.

-The **vitreous humour**-Which is under turgor pressure. It helps to maintain the shape of the posterior chamber of the eye ball. It also plays an important part in the refraction of light rays enabling them to be focused on the retina.

-The cornea, transparent and curved which helps to play an important role in focusing of the image on the retina. It accounts for the largest refraction of light rays.

-The aqueous humour –Contains oxygen and nutrients, which nourish the cornea and the lens. It is under pressure thus helping to maintain the shape of the anterior chambers of the eye. It also plays a part in the refraction of light rays enabling them to be focused on the retina.

-The Iris is heavily pigment, to prevent entry of light into the eye except through its central aperture called the pupil. It contains circular and radial muscles which constrict or dilate the pupil depending on the intensity of light.

-The lens is elastic, therefore allows changes in its shape depending on the tension exerted through the suspensory ligaments. This enables it to bring light rays coming from either near or far objects into sharp focus on the forea.

-The ciliary's body Contains the ciliary muscles whose contraction and relaxation alters the tension exerted on the suspensory ligaments.

-This in turn alters the shape of the lens enabling it to focus for both near and distant objects.

-The eyelids which are movable and opaque structures can be closed through a reflex action to protect the eye from too much light or from foreign objects.

- **The eye muscles** help to move the eye ball within the orbit. The lateral

  rectus muscles move the eye up and down whole the oblique

muscles, move the eyeball in its up and down movement.

-The lachrymal gland which continuously secretes a watery, saline and

antiseptic fluid called tears. The tears moisten the cornea and wash

foreign particles out of the eye.

-**The eyelashes,** which are many hairs, protect the eye from the entry

of small foreign particles.

-**The eyebrows** raised portion of the skin above the eye, thickly covered

with hair, whose functions are to prevent sweat and dust from entering the eye.

EXERCISE THREE

1. The photograph below sh.... a tooth from a carnivore and tooth fi..... a herbivore.

A

B

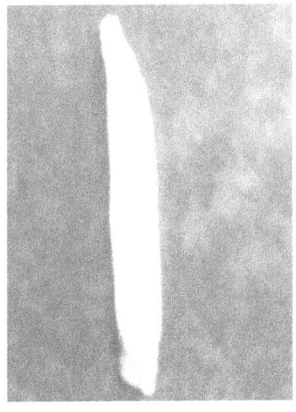
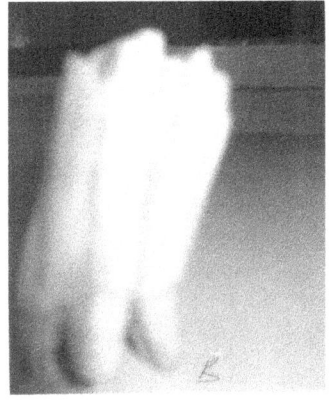

(a) Which   tooth

is   from a

carnivore and which tooth is from a herbivore?

From a carnivore -

Reason -

From a herbivore –

Reason - .

(b) Identify the type of teeth shown by the photograph B.
Type -

Reason -

(c) State the functions of the tooth B shown. Give reasons for your answer.
Functions -

.

Reasons -

(d)   Distinguish between homodont and heterodont teeth.

(e)   Explain any two general adaptations of tooth to its functions.

2.   The diagram below shows part of a mammalian skeleton, Study it and use it to   answer the questions that follow.

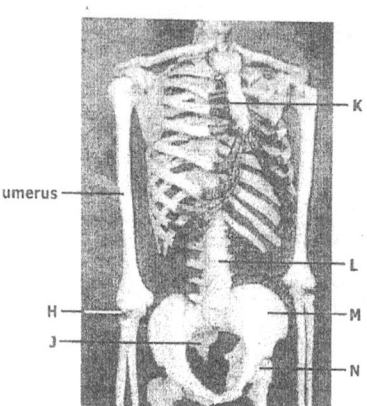

(a)   Name each of the parts of the skeleton marked H, J, K and N.

(b)   Name each of the parts of the human skeleton described below.
        (i) The part on which the anterior portion of N articulates.

        (ii) The three bones that together fuse to form bone M.

(c)   State any **two** adaptations of each of the following structures:
        (i) Structure M

        (ii) Structure L

(d)   On the diagram label each of the following parts using the letters in brackets.
        (i) the pubis symphysis (P).
        (ii) the part where intercostals muscles attach (I)
a joint that can turn through 180^0 only.

ANSWERS TO EXERCISE THREE

A

B

1. The photograph below sh...... a tooth from a carnivore and tooth from a herbivore.

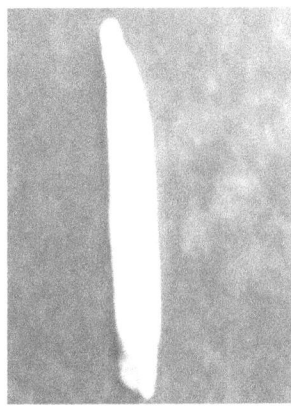
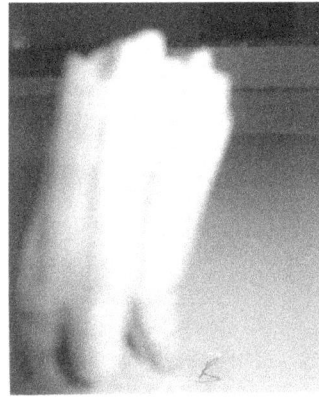

(a) Which tooth

is from a

carnivore and which tooth is from a herbivore?

        From a carnivore - *A;*

        Reason - Sharp and pointed for tearing flesh from bones;

        From a herbivore – *B;*

        Reason - *Has cusp for chewing and grinding;*

(b)   Identify the type of teeth shown by the photograph B.

        Type - *Pre-Molar or Molar; any one*

        Reason - *Has three roots;*
              - *Large surface for chewing and grinding*

(c)   State the functions of the tooth B shown. Give reasons for your answer.

        Functions -

           Chewing and grinding; of vegetation and grass.

        Reasons -

Sharp cusps; and wide surface for chewing and grinding;

(d)   Distinguish between homodont and heterodont teeth.
        Homodont type are similar in shape and size while heterodont are
        different in size and shape.

(e)   Explain any two general adaptations of tooth to its functions.
        Long to pierce through flesh;
        sharp to tear flesh from bones;

2.   The diagram below shows part of a mammalian skeleton, Study it and use it to   answer
the questions that follow.

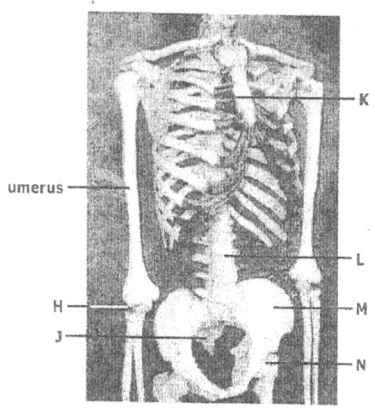

(a)   Name each of the parts of the skeleton marked H, J, K and N.
        H   -   Hinge joint

        J   -   Coccyx

        K   -   Cartilage of rib / rib;

        N   -   Head of femur

(b)   Name each of the parts of the human skeleton described below.
        (i) The part on which the anterior portion of N articulates.
            Acetabulum;

        (ii) The three bones that together fuse to form bone M.

Ilium; ischium; and pubis;

(c)   State any **two** adaptations of each of the following structures:
   (i) Structure M
        Has large surface area for attach of high muscles; has a cetabulum
        which articulates with head of femur.

   (ii) Structure L
        Large and broad centrum to offer support;

        Broad and long transverse processes; for muscle attachment

(d)   On the diagram label each of the following parts using the letters in brackets.
   (i) the pubis symphysis (P).
   (ii) the part where intercostals muscles attach (I)

        (iii)a joint that can turn through 180^0 only.

EXERCISE FOUR

1. The scientific name for French bean is *Pharseolus vulgaris*
   (a) What taxon does the term Phaseolus represents?
   (b) State **two** rules that are followed when giving a scientific name to an organism.

2. a) What is the function of the mirror in the microscope?
   b) Which organelle would be abundant in:
   Skeletal muscle cell
   Palisade cell

3. A seedling shoot was exposed to unidirectional light as shown below. The set up was left in the dark room for three days.

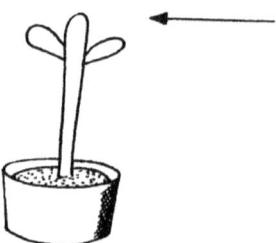

   i) Make a drawing of the expected results at the end of the experiment.
   ii) Explain the expected results at the end of experiment.

4. State **two** advantages of breathing through the nose than through the mouth.

5. Name **two** mineral elements required in the synthesis of chlorophyll.

6. a) State **two** environmental condition that can cause seed dormancy.
   b) Name the part of the leaf that elongates to bring about epigeal germination.

7. a) State the function of amylase in human body.

    b) Name **two** parts of the alimentary canal where amylase is secreted.

8. a) Name **two** photochemical cells in human retina.

    b) Name **one** chemical substance and two mineral ions involved in impulse transmission in mammals.

9. Give the function of melanin pigment produced in the skin of man.

10. What is the importance of saprophytic bacteria in an ecosystem?

11. A student while carrying out an experiment observed 8 cells across the field of view of light microscope. If the diameter of the field of view is 5 mm, calculate the average length of each cell in micrometers.

12. State **one** feature present in the flowers that can be used to distinguish between a monocotyledonous flower and dicotyledonous flower.

13. The graph below shows levels of oestrogens and progesterone during the human menstrual cycle.

    a) Mark on the graph the curve that represents

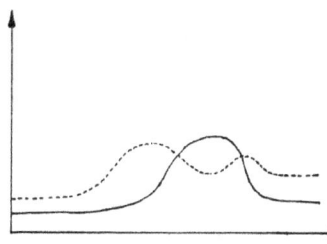

i) Progesterone

ii) Oestrogen

b) Which is the most likely day of ovulation from the graph?

14. a) What are fossils?

b) State **two** limitations of the use of fossils as an evidence of evolution.

15. Name the type of skeleton in:

i) Grasshopper

ii) Sheep

16. Name the type of response shown by;

a) Leaves of *Mimosa pudica* when they fold after being touched.

b) Sperms when they swim towards ovum

c) Euglena when they swim towards the source of light.

17 a) Give an example of sex linked trait on x-chromosome.

b) Below is a nucleotide strand.

A	A	G	T	C

i) Identify the type of nucleoic acid strand.

ii) Give your reason for your answer in (b) (i) above.

iii) Write down the complimentary base sequence in the other strand.

18. The diagram below shows a stage in cell division

i) Name the stage of the cell division that exhibits the process above.

ii) What is the significance of the phenomenon shown to a species?

19. Differentiate between respiration and respiratory surface.

20. State **two** adaptations of skin of the frog to gaseous exchange.

    a) A man's urine gave a positive reaction with Benedict's solution. Name the disease he was suffering from.

    b) State **two** ways in which the symptoms of the condition in (a) can be controlled.

22. A student collected an organism in the school compound and noted it had a segmented body and two   pairs of legs per body segment.

    i) Name the class to which the organism belongs.

    ii) State **two** other features the student may have observed.

23. a) Name **two** structures of gaseous exchange in aquatic plants.

    b) What is the effect of contraction of the diaphragm muscles during breathing in mammals?

24. The diagram below represents part of the mammalian blood circulatory system and some associated glands.

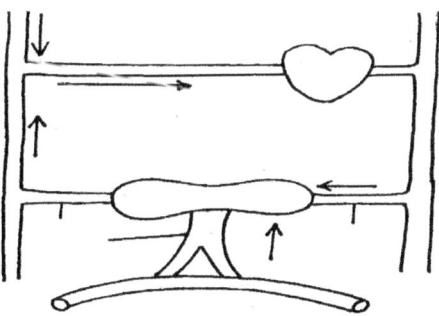

    (a) Name the blood vessels **A** and **B**

    (b) State **two** structural differences between the blood vessels labelled **A** and **C**

25.   A student made equidistant marks on a radical of a dicotyledonous seedling. After three days the distance between the marks was measured.

a) What was the aim of the experiment?

b) Predict the results that were likely to be obtained by the student

26.   a) Name the disease caused by H.I.V

b) Give **two** reason why it is difficult to cure the disease named above.

c) Give **one** preventive measure of the named disease.

27.   Plants of a particular species grown in certain habitat flower at the same time. What is the importance of this adaptation

28.   State **two** roles played by the bark in plants

29.   The diagram below represents a bone obtained from a mammal.

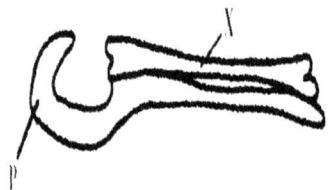

i) Name bone labeled **X**.

ii) Name structure **P**.

30.   A student mashed a piece of ripe banana and made it into paste by adding water, place the paste in a visking tubing and suspended it in a beaker containing iodine solution as shown below. The set up was left for 40 minutes.

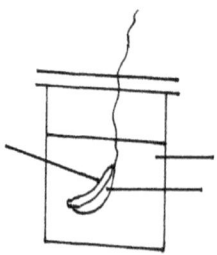

a) State the physiological process under investigation.

b) Account for the result obtained in the table.

31.   Industrial waste may contain metallic pollutants. Explain how the pollutants may indirectly reach and accumulate in the human body when the wastes are dumped into rivers.

32.   During oxidation of certain foods substances the respiratory quotient was found to be 0.718.

i) Name the type of food substance being oxidized.

ii) State **two** advantages of using the food substances named.

ANSWERS TO EXERCISE FOUR

1. a) genus

    b) – the genus name should be in capital letter and the species name with a small letter:
    - should be printed in italic or when ad written should be underlined as separate

words:

2   – should be Latinized i.e. made to sound like Latin words: (2 correct responses 1x

2mks)

        a) Reflects light through the source through the condenser to the stage:
            (1)

    b) - Mitochondria
       - Chloroplasts

3.   (i)

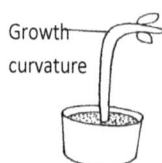

Growth
curvature

    (ii) The shoot tip bends towards the direction of the light: auxins diffuse to the shaded
    size causing more growth than on the side receiving the light:

4.   - Air is cleaned cilia.
    -    Amount of air taken in is controlled.
    -    Any smell in the air is detected

5.   - Magnesium
      - Nitrogen:
      - Iron. Acc. Magnesium ion/ iron rej symbols of elements

6.   a) – absence
        - Unsuitable temperature
        - Lack of Oxygen

    b) Hypocotyls

7.   a)   Digests starch to maltase
      b)   -   Mouth
            -    Duodenum:

8.   a)   Rods;   cones
      b)   Chemicals; acetylcholine/ noradrenalin
            Mineral ions: Sodium; Potassium; chloride ions

9.   -   Screen or protects against ultra violet rays from the sun

10.   -   Decompose organic matter/ recycle organic materials:

11.   Average length of one cell = field diameter/ no. of cells
       5mm/8t: 0.625mm
       0.625 x 1000 = 625 micrometers

12.   The floral part in monocots are in three or multiples of three.
       In dicots they are in fours. Fives or their multiples: *(mark as a whole)*

13.   a) i) ....Oestrogen hormone: √

ii) _____Progesterone hormone:

b) The 14th day;

a) Fossils are past materials remains of ancestral forms of organisms that were accidentally preserved in natural occurring materials;   (OWTTE)

b) - Only partial preservation was possible due to soft parts decayed;

- Distortion during sedimentation;
- A subsequent geological activities(erosion, earthquakes, faulting) destroyed some fossils;
- Missing fossil records/ missing links

15.   i)   Exoskeleton

ii)   Endoskeleton

16   a)   nastic/ naptonasty / thigmonasty;

b)   Chemotaxis;

c)   Phototaxis

17   a)   Haemophilia;/colour blindness:/ Eye colour in drosphila;

b) i) DNA

ii) Has nitrogenous base Thymine

iii)

T	T	C	A	G

18. i)   - Prophase 1;

ii)   - It leads to genetic exchange that brings about variations;

19.   - Respiration is the chemical breakdown of glucose to release energy

- Respiratory surface is the surface across which respiratory gases exchange

20. Moist for gases to dissolve efficient diffusion;

- Large to provide a large surface area over which gaseous exchange takes place;
- Highly vasculatrized for fast transport of gases in and out of the ski;
- Thin epithelium for fast diffusion of gases;

21.   a)   Diabetes militus

b)   Regular injection with insulin

Diet – reduction in the amount of carbohydrates taken in;

22   i) - Diplopoda;

ii) – Cylindrical body

- Three body parts;
- A part of short antennae
- Two clumps of many simple eyes;

23. a) -Pneumatophores

- Aeronchyma tissues
- Cuticle

b) - Eternal intercostals muscles relax while internal intercostals muscles contract; moving the rib cage

Downwards and inwards: The diaphragm muscles relax making it dome shape;

- The volume of the thoracic cavity decreases while its pressure increases pushing air outside due to low atmospheric pressure;

24. a) A- Hepatic portal vein;   B – Hepatic vein

b)

Hepatic portal veins	Hepatic artery
- Wide/ large lumen	-Narrow lumen
- Presence of valves	- Absence of valves
Thinner walls/ less muscular walls	- Thick walls/ muscular walls

25.   a) To investigate the region of elongation in roots;

b) the distance between the marks just behind the root tip increased; this is the religion cell elongation there cells expand and increase in length;

- few glomerulli;

26. a)   Acquired Immuno – deficiency Syndrome/ AIDS

b) It destroys the immune system virus replicates rapidly

Virus is obligate intracellular

c)   - Screen blood for HIV before transfusions;

- Avoid multiple sexual partner
- Sterilize surgical implements before use
- HIV positive mothers should avoid breast feeding

27.   -Encourages cross pollination and fertilization/ Hinders self pollination and fertilization;

28.   – Insulation against fire;

- Protects against infection by fungi;
- Prevents damage by insects
- Prevents loss of water;

29.   i) Radius;√1   (ii) Olewanon process; √1   (iii) Humerus; √1   *rej.* Humerous

30.   a) Diffusion

b) Iodine molecules from the beaker moved into the visking tubing by diffusion since they were small in size; (iodine reacted with starch to form blue- black colour). Colour of iodine solution outside remain brown as starch molecule were too large to pass through the visking tubing.

31.   - Aquatic plants absorb the metallic ions through their roots and are incorporated in there tissues: when eaten by fish the pollutants become incorporate in fish tissues; such fish when eaten by humans, accumulate in the tissues;

32. (i) Fats/ lipids – accumulated ions

(ii) Release large amount of energy per molecule on complete oxidation; √ produces a lot of water on oxidation; √ (use for other metabolic activities)

EXERCISE FIVE

1. The diagram below represents part of a cockroach gaseous exchange system.

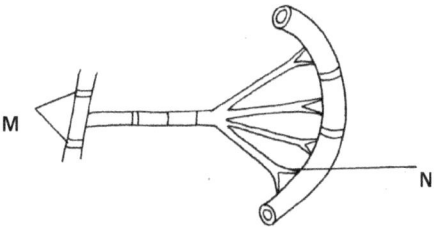

M

N

a) State the function of the part labeled **M**

b) Suggest how the part M is adapted to the gaseous exchange function

c) How does the movement of oxygen in an insect and mammals from atmosphere to the tissue

of its body differ

2. The following chart below shows blood transfusion pathway

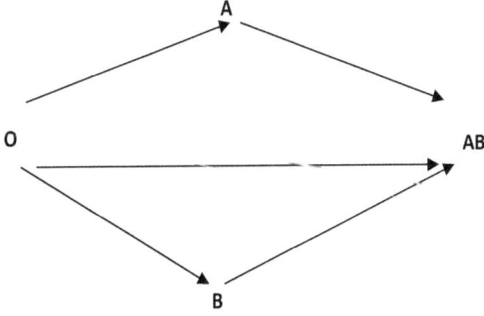

A

O

AB

B

a) What **five** conclusions can you draw from the flow chart

b) Why is the knowledge of blood groups necessary before blood transfusion?

c) A part from knowledge of blood groups, state two precautions that must be observed during

blood transfusion

3. The genetic disorder haemophilia is due to a recessive sex linked gene. A man who is haemophiliac married a woman who is a carrier for the condition.

a) Using letter (H) to represent normal condition and (h) to represent haemophiliac condition.

i) What is the genotype of the man and the woman?

Man _____

woman_____

ii) Work out across between the man and the woman

b) What are the chances that both the first and the second sons will be haemophiliac?

c) Haemophilia is most common in the males than females humans. Explain.

4. The diagram below shows different groups of organisms and their biomass.

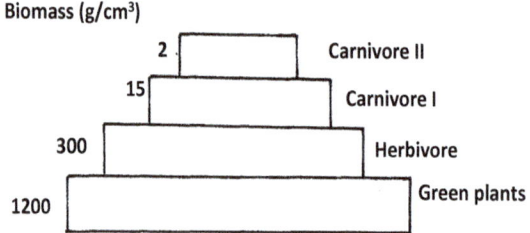

Biomass (g/cm³)

2 — Carnivore II
15 — Carnivore I
300 — Herbivore
1200 — Green plants

a) Define the term biomass

b) Account for the decrease in biomass in the successive group of organisms

c) Describe how energy from the sun is made available for carnivore II

5. Cell of a certain herbaceous plant were found to have an average diameter of 2.5µm the cells were put in   varying concentrations of salt solutions. The average diameter of the cells in each solution was determined and the results were recorder as shown in the table below.

Concentration of salt solution %	Diameter of cells. µm
1	5.0

5	4.0
10	3.0
15	2.0

a) From the results above, determine the cell sap concentration

b) Give an explanation for the average diameter of the cells placed in the following salt concentration compared to the normal diameter of the cells.

i) 1 % salt solution

ii) 15 % salt solution

Give the term used to describe salt solution whose concentration is the same as cell sap.

6. In the experiment, the population growth of yeast cells in a Petri dish was determined over a period of 75 minutes. The results below were obtained.

Time in minutes	Number of yeast cells
0	4
5	6
10	8
15	10
25	30
30	50
35	80
40	120
45	140
50	150
55	160
65	166

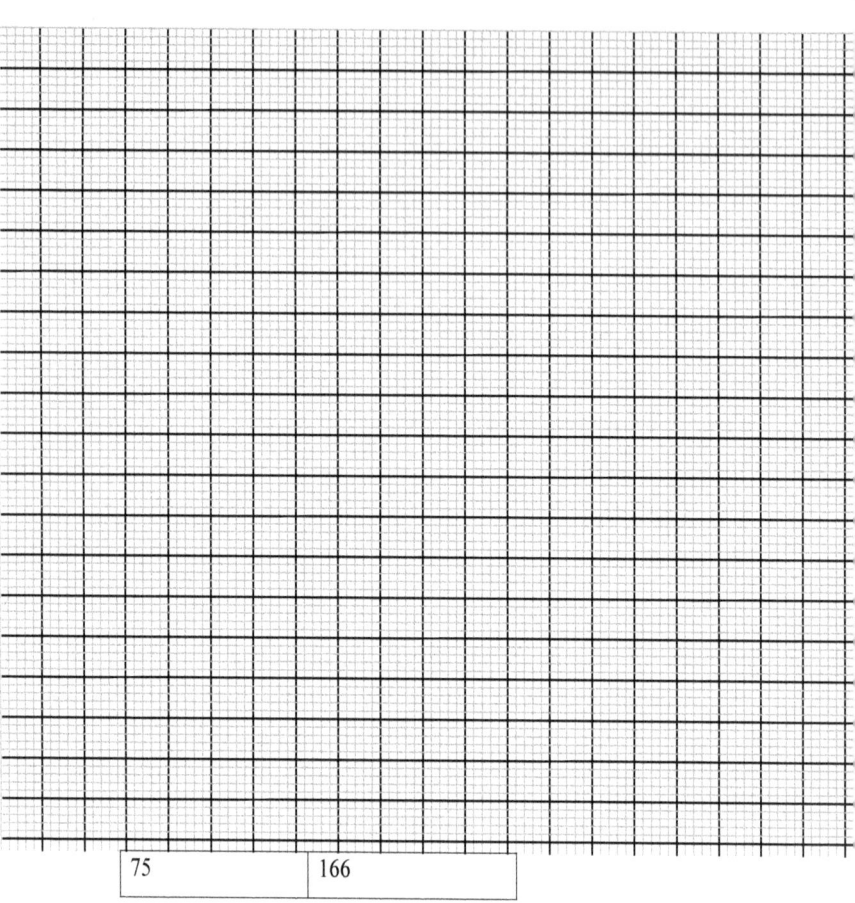

75	166

a) Using a suitable scale, plot a graph of number of cells against time in minutes

b) Name the type of the curve you have drawn above

c) Determine the number of yeast cells after 37 minutes

d) After how long was the population of yeast cells 144?

e) Work out the rate of cell division between 32 minute and 42 minute

f) Account for the shape of graph between 45th minute and 60th minute

g) In a field study to estimate the population of grasshoppers in the school field of 4 km^2, 60 grasshoppers were caught using sweep nets, marked with red paint and released back to the field. The following day students went back with their sweep nets and caught 100 grasshoppers, in which 20 were found to be already marked.

i) Calculate the population size of grasshoppers in the field

ii) Calculate the population density of the grasshoppers in the field

iii) What factors would maintain the population of grasshoppers and yeast cells at the carrying capacity.

7.   Describe the various evidences to support organic evolution

8.   a) Describe how the heart beat is controlled and maintained

b) Describe the structure and function of thrombocytes

ANSWERS TO EXERCISE FIVE

1. a) Strengthen the wind pipe;
   b)   They ramify the body tissue for direct supply of individual cell with oxygen;
          - They have thin walls for easy penetration of gases;
          - They have moist cell wall that dissolves oxygen;

c) Insect	Mammals
- Transported in the tracheal system	-transported in blood vessels
- Transport fluid in coelom	- transport fluid in blood
- entry in thro' the spiracles	- entry is thro' the nose

2 a)– Blood group AB individual can receive blood from individuals of all other blood groups
   -     Blood group AB individual can only donate blood to individuals of blood group AB
   -     Blood group O individuals can donate blood to all the other groups
   -     Blood group A individual can donate blood to group A and AB individuals
   -     Blood group B individuals can donate blood to blood group B and AB individuals
   b) Because if blood of individual with different antigen mix agglutination; of the blood will occur in   blood vessels leading to blockage of vessels and later death;
   c) The donor must be healthy;
   The donor must be between 18 – 65 yrs of age;

3. a) i)   Man- X^hY;
             Woman- X^HX^h;

      ii)

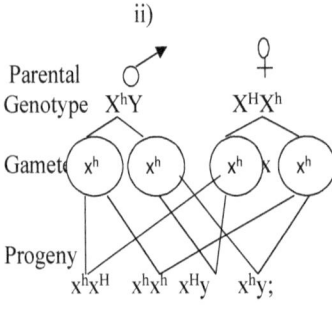

Parental
Genotype X^hY X^HX^h

Gamete x^h x^h x^h x^h

Progeny
         X^hX^H X^hX^h X^Hy X^hy;

   b)   1/2 or 50 %;
         1/2 or 50 %;
   c)   Because the **Y** chromosome lacks the genes that control the trait; such that any time the **X** that is   donated by the mother has the recessive allele (h) there is no (H) on the **Y** from the father to mask the boy.
4. a) Biomass is a constant; dry weight of an organism;
   b)   Energy is lost in respiration;

- Primary producers source energy directly from the sun;
- Loss through defaecation;

c) green plants absorb the energy from the sun during photosynthesis; then herbivores get it when they   feed on green vegetation; its acquired by carnivore I as it preys on the herbivores and then to carnivore II as it preys on the carnivore I;

5. a) Midpoint between the diameter between the diameter of 3.0 μm and 2.0 μm

$$= \frac{10+5}{2} = \frac{25}{2} = 12.5\%;$$

b) i) 1% salt solution in hypertonic solution to the cell sap; an osmotic gradient is created between the cell sap and salt solution; making water molecule to be drawn into the cell by osmosis; hence increases in diameter as the cell become turgid.

ii) 15% salt solution was hypertonic to the cellsap; an osmotic gradient was created between the cell sap and salt solution; making water molecules drawn out of the cells by osmosis; hence   the cell become flaccid and decreased in the diameter.

c) Isotonic;

6 a) Ploting; scale vertical; horizontal; smooth curve; Axes; label

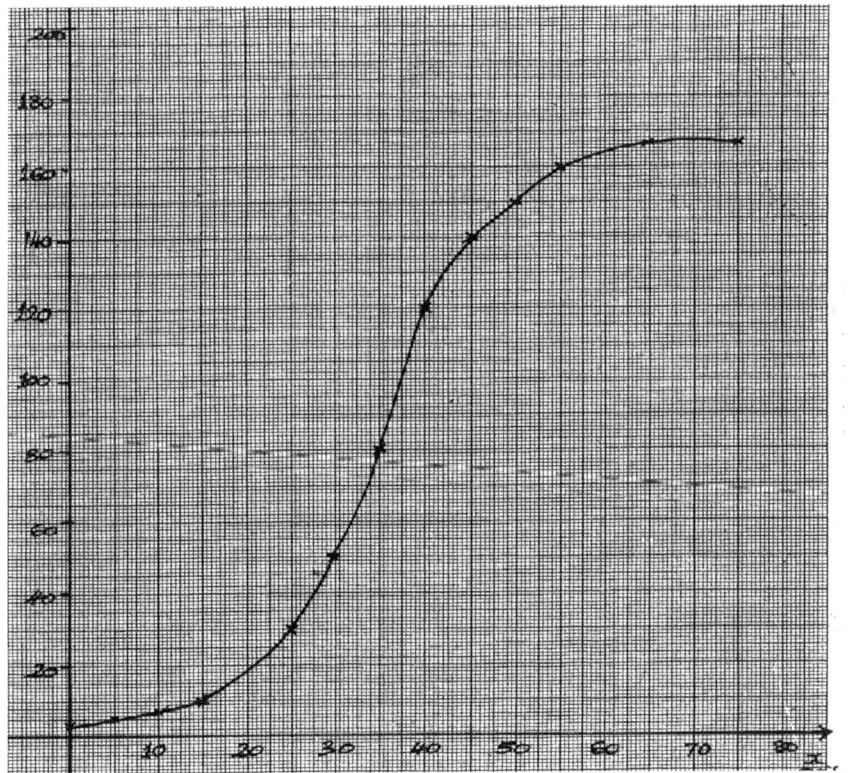

b) Sigmoid curve;
  c) 98 yeast cells +- 1;
  d) 47 minutes;
  e) Rate of cell division = Change in minutes over
                              Change in time

$$= \frac{128 - 60}{42 - 32}$$

$$= \frac{68}{10} \quad = 7 \text{ cells/ minutes}$$

f) Rate of cell division is decreasing with the increase in time; due to shortage of oxygen and nutrients;
Space is limited, accumulation of metabolic wastes which inhibits multiplication;
  g) i) population = First marked x second capture
                      Marked capture

$$\frac{100 \times 60}{20}$$
    = 300 grasshoppers;
  ii) Population density = Total population
                            Area
$$= \frac{300}{4}$$
    = 75 grasshoppers/ km^2
  iii) Competion;
Death of those not suitable adapted;
7. – **Fossil evidence / paleontology**; fossils provide direct evidence of evolution; the relationships between extinct organism and existing ones; is shown by similarities between skeletons; hard parts and the skeleton structure of the species that are in the existence; e.g. Homo erectus, homo habilis have similar skeleton or insect preserved have similar structures to existing ones.
- **Geographical distribution;** apparently animals of common origin occupy similar geographical locations in different continents; e.g. camels in Africa and Llama in South America are found in the same latitude; or leopard, Tiger, and Jaguars are of common origin and occupy similar geographical locations in different continents i.e. in Africa, Asia and America respectively, this distribution is due to "continental drift";
- **Comparative embryology**; unrelated organisms in different classes of vertebrates have similar embryonic developmental patterns and structure; e.g. gill slits/ clefts are present in every early embryonic stages of mammals, Pisces, amphibian and aves; or vertebrates have a notochord at least at one developmental stage suggesting common ancestry.
- comparative cell physiology / Biochemistry/ serology; Analysis of blood proteins of unrelated animals reveal similar contents e.g. antigen – antibody reaction suggests common ancestry; if human serum is injected into a rabbit, secrete antibodies against human antigen;

- **comparative anatomy**; unrelated organisms have similar anatomical structures; pentadactyl limbs of mammals, bird, reptiles suggests common origin; these structures look different from others due to divergent evolution thus produces homologous structures; also the caecum in rabbits has developed due to use while appendix has become vestigial due to disuse;

8 a) The sino autrial node initiates and maintains the heartbeat; by generating a wave of electrical signals that spreads through both atria; making them contract simultaneously; the signal then spreads to the autria-ventricular node (AVN); during which the atria empty into the ventricles; the signals spreads to the purkinje fibres; then conduct signals to the apex of the heart; and through the ventricular walls; these signals triggers a wave of powerful contraction of both ventricles; from the apex towards the atria driving blood in large arteries; the cardiac muscles are myogenic hence not controlled by nervous stimulation.
Any ten correct marking points x 1 = 10 marks

b) Thrombocytes are blood fragments that are irregularly shaped; they lacked a nuclear and they play a   major role in blood clotting process.

When a damaged blood vessel is exposed to air; the inactive enzyme prothrombin is converted to active enzyme thrombin; under influence of thromboplastin factors like Ca^{2+}; thrombin the converts soluble plasma proteins fibrinogen; into insoluble protein fibres fibrin; fibrin forms a fine mesh over the wound trapping blood cells; and large proteins to form a soft fibrin clot; serum oozes out through the clot; and due to exposure to air it dries up and hardens to form a scab; which serves to protect soft underlying tissue and allow it to heal quickly.

www.ingramcontent.com/pod-product-compliance
Lightning Source LLC
Chambersburg PA
CBHW071006180526
45168CB00003B/1310